Motorsport

Racing

Crossword Puzzles

Danielle McCorkle

Please visit the websites listed in the Sources page for more great reading about Motorsport Racing.

Table of Contents

Types of Racing Puzzle 1

ACROSS

2 Race Car looks similar to a road car, usually races on an oval track
5 Sometimes youth start with this type of racing, often the first step in racing
7 National Association for Stock Car Auto Racing LLC
10 Open wheeled auto racing with speeds up to 240 mph
11 Racing where the competition is two cars racing each other
13 International racing with open wheeled racing cars
16 Competing on an uphill track

DOWN

1 Modified race cars or stock cars race on a dirt track
3 Racing using cars designed for regular roads, cars may be modified for racing
4 Racing on highways that have been closed for the purpose of this race
6 A race over grass and hard surface road with many cars competing
8 Racing using cars with two seats
9 A series of races in one season with open wheeled cars
12 Racing using modified vehicles
14 A variety of races with different competitive events
15 Electric vehicle racing

Dirt Track Racing	Indy	Rallying
Drag Racing	Indy Car Series	Sports Car Racing
Formula E	Kart Racing	Stock Car Racing
Formula One	NASCAR	Touring Car Racing
Grand Prix	Off Road Racing	
Hill Climb Racing	Rally Cross	

Types of Racing Puzzle 1

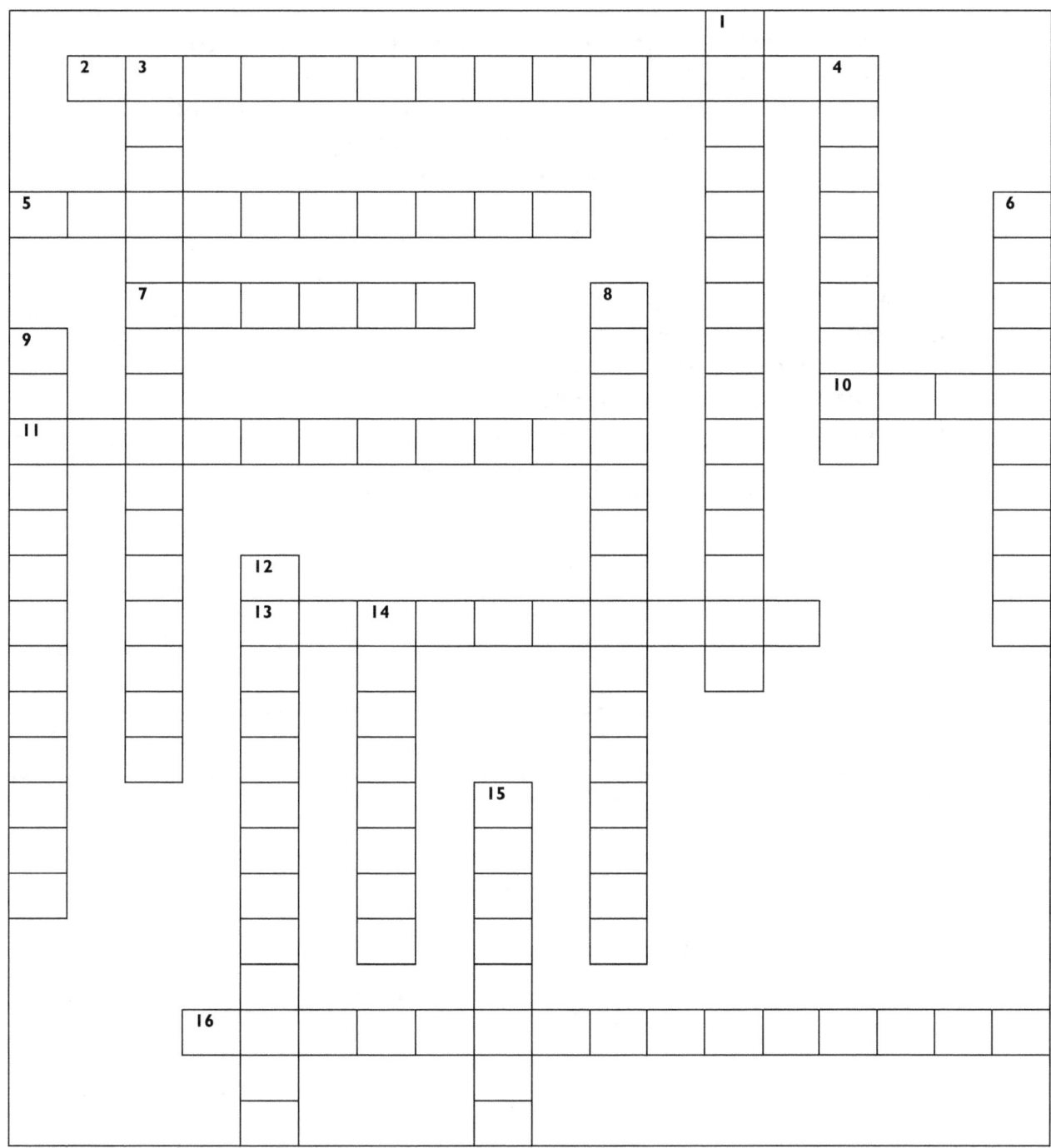

Races Puzzle 2

ACROSS

1 The National Association for Stock Car Auto Racing series of races
6 The United States Auto Club
8 The NASCAR race series that involves stock cars, called the second-tier
12 This 500-mile race is held Memorial Day Weekend, named the _____ 500
13 National Hot Rod Association racing

DOWN

2 An annual race in Colorado USA
3 A pickup truck series of racing
4 This race originated in France
5 A 24-hour endurance race
7 An American race with open wheeled cars
9 International racing with open wheeled, single seat cars
10 Extreme dirt motorsport racing on multiple surfaced courses
11 A 500-mile race held in Florida, named the _____ 500

Cup Series	Le Mans Series	NHRA
Daytona	Indianapolis	Rally Cross
Formula One	Indy Car	USAC
Grand Prix	NASCAR Truck Series	Xfinity Series
Pikes Peak International		

Races Puzzle 2

Racing Facts Puzzle 3

ACROSS

3 The Indianapolis Motor Speedway's nickname
4 Formula One racing began here
6 The person who holds a sign telling the driver to go to the pit
7 The first Daytona 500 raced here
8 The term for the driver who get the best starting spot for the race
10 The original road material at the Indianapolis Motor Speedway
13 The meaning of Grand Prix in French
14 The year the first Indianapolis 500 was held

DOWN

1 A tradition of the Indy 500, given to the winner
2 This determines the line up of the cars prior to the start of the race
4 The checkered flag signifies this
5 Wheels located outside of the body of the car
9 The cars line up here to begin the race
11 Number of laps around the track for the Indianapolis 500
12 The area where the race car driver sits

A Bottle Of Milk	End Of The Race	Open Wheeled Cars
A Pole Sitter	Europe	Qualifying Rounds
Bricks	Grand Prize	Racing On The Sands
Brickyard	Lollipop Man	Starting Grid
Cockpit	Nineteen Eleven	Two Hundred

Racing Facts Puzzle 3

More Racing Facts Puzzle 4

ACROSS

1 Walls to protect the racer and spectators
6 This signifies Caution
7 This signifies One More Lap
8 This signifies Go
9 Where the cars line up for the race
10 Where the winner drives his car to after winning the race
11 The place during the race where the driver can get more fuel, new tires etc
12 The number one spot every driver would like to be at
13 This signifies a Warning
14 What is in place for crash prevention at the race tracks

DOWN

2 This signifies Stop
3 The year racing officially began
4 This signifies the Winner
5 This signifies Pick up the Pace
7 This person started Motor Magazine in 1903

Barriers	Green Flag	Victory Lane
Black Flag	On The Pole	Wall Crash Prevent
Blue Flag	Pit Stop	White Flag
Checkered Flag	Red Flag	Yellow Flag
Eighteen Ninety-Five	Starting Grid	
William Randolph Hearst		

More Racing Facts Puzzle 4

Parts of Race Cars Puzzle 5

ACROSS

2 Protects the driver in case of a roll over
7 Race cars are aerodynamically designed because of this
11 This is used to cool down the engine
14 When the tires are under the race car's body
16 This button allows the driver to shift gears without touching the gear level

DOWN

1 These are the designers of race cars
2 These limit the speed of the race car
3 Protects the head during an accident
4 This has a roof
5 This is how a driver enters the car
6 The gas tank in a race car
7 Some of these are made with shatterproof plastic
8 Padding used to protect the driver
9 The body of the race car
10 AFP system is used to deflect debris during racing, Advanced ____ ____
12 This has no roof
13 This is inside the tire in case it goes flat during the race
15 When the tires are outside of the race car's body

Airflow	Fuel Cell	Restrictor Plates
Chassis	Head Restraint	Roll Cage
Closed Cockpit	Inner Liner	Stock Car Window
Closed Wheels	Open Cockpit	Thick Foam
Fabricators	Open Wheels	Wind Resistance
Frontal Protection	Paddle Shift	Windshield

Parts of Race Cars Puzzle 5

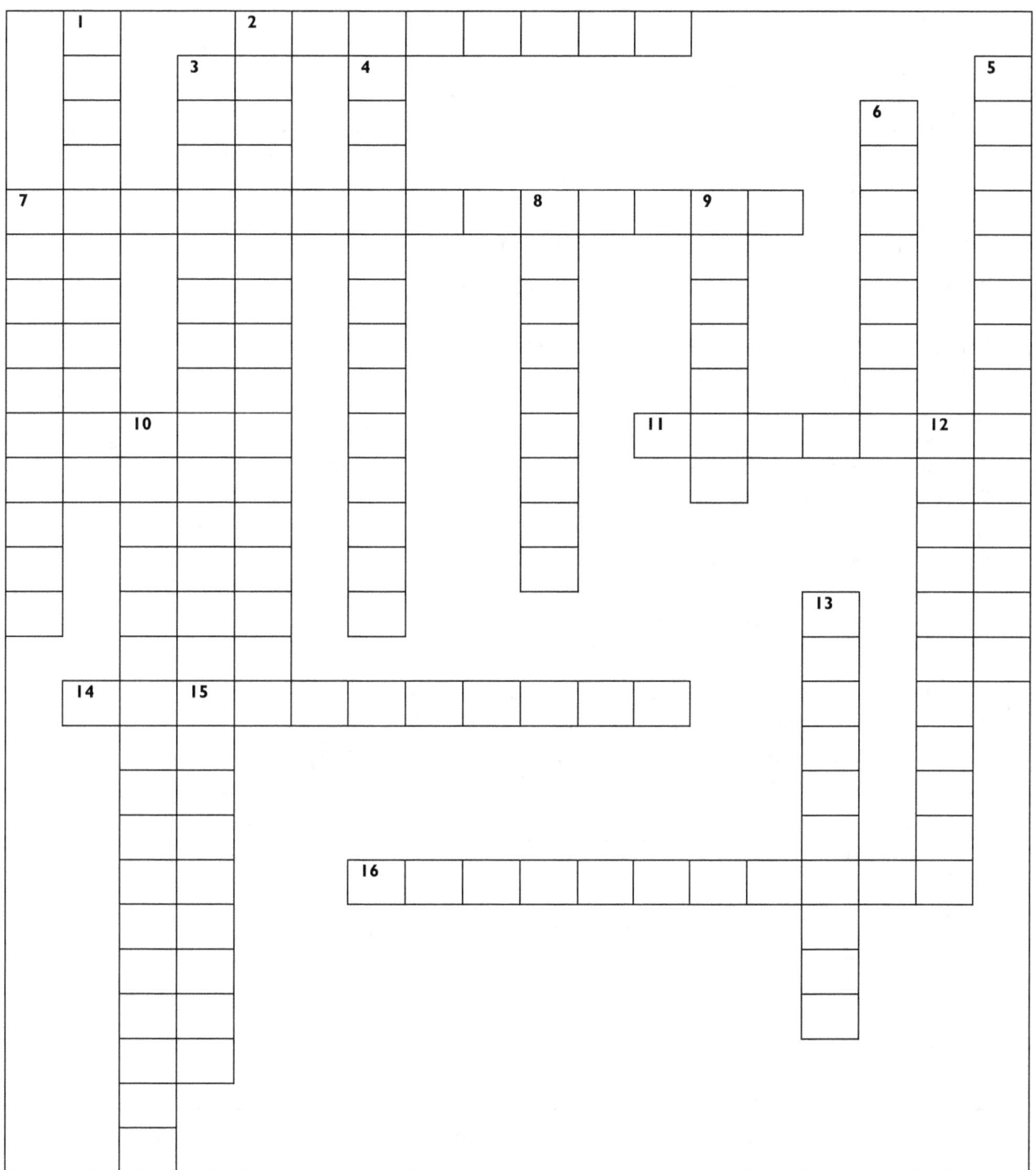

Pit Crew Puzzle 6

ACROSS

1 On the Pit Road, this is in place
4 The person who operates the air jack
8 The leader
10 Where the Pit Crew goes to keep up their strength and conditioning
11 Where cars go to get serviced during the race
14 Part of the Pit Crew Staff
15 The Pit Crew must wear this when working
16 There are usually 4 of these in the Pit Box
17 This is timed when working
18 There are many of these to watch out for
19 The Pit Crew's clothing must be this

DOWN

1 This is what the Pit Crew practices and follows it in order to win
2 This person in the Pit Crew catches fuel spills
3 This is important for those in the Pit Crew
5 Part of the Pit Crew Staff
6 Sometimes call the "War Wagon"
7 The Pit Crew must wear this
9 Fills Fuel Tanks
10 The Pit Crew must wear to protect their hands
11 Certain schools offer this program
12 If the Pit Crew is not careful, this could start where they are working
13 This could happen if you touch hot tires if you don't wear gloves

Approved Safety Gear	Gloves	NonFlamable
Burned Hands	Gym	Pit Box
Catch Can	Hand Speed	Pit Crew Training
Crew Chief	Hazards	Pit Stall
Engineers	Helmets	Speed Limit
Fill Fuel Tank	Jackman	Strategy
Fire	Mechanics	Tire Changers
Hand Eye Coordination		

Pit Crew Puzzle 6

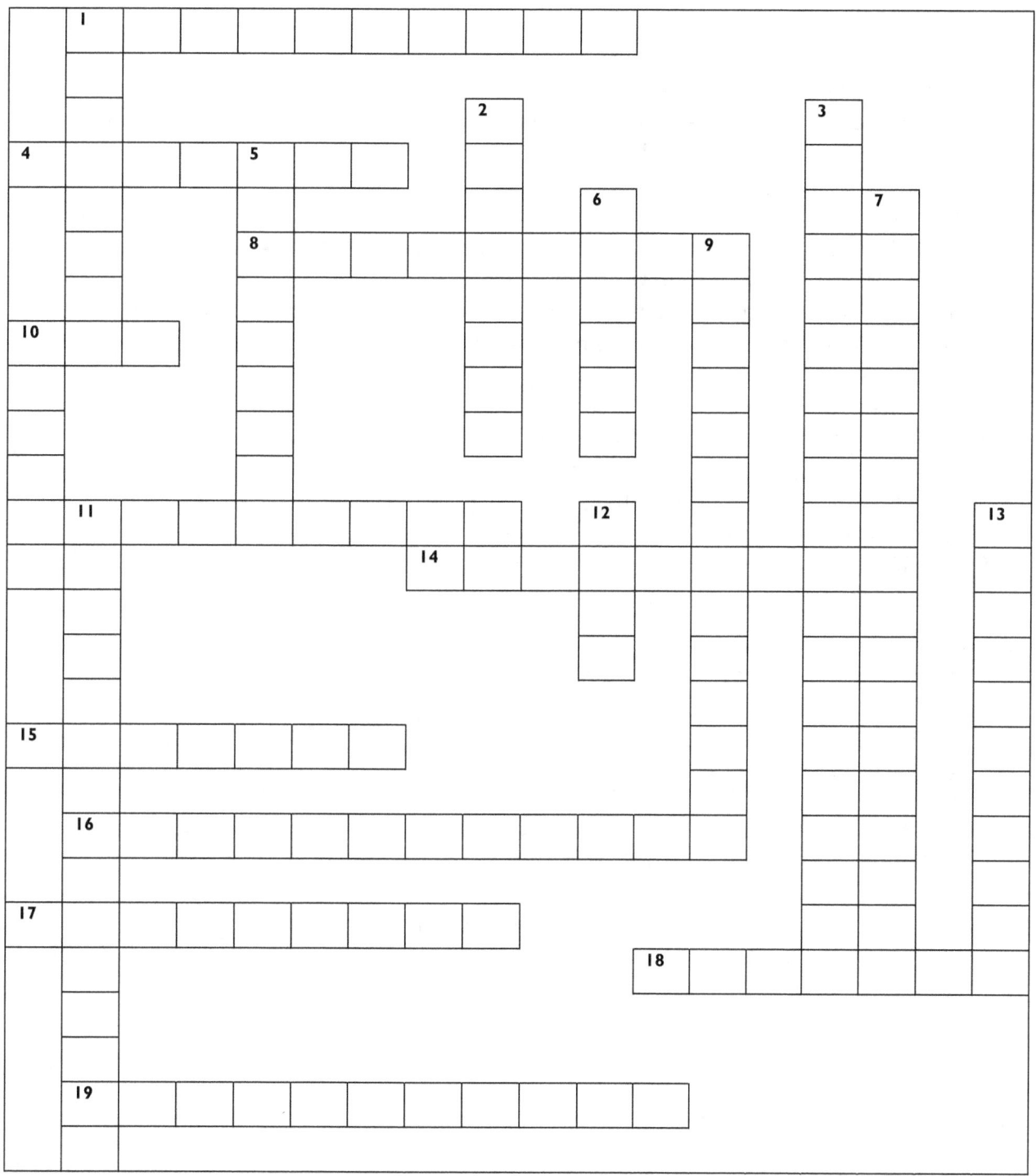

15

Some Differences Between Race Cars Puzzle 7

ACROSS
1 Open wheeled cars have this type of engine
3 Most of the NASCAR races have this shaped track
6 Enclosed wheeled cars have this weight of car vs open wheeled race cars
7 Open cockpit cars have this weight of car vs closed cockpit race cars
9 Race cars with the tires below the body
10 Classic position for drivers in open cockpit cars
11 Indy Series Racing has less of this than other series each year
12 How drivers get into a closed cockpit car
13 Racing by Indy Car allows this which can knock the car out of the race
14 When the cars are lighter, they are usually this also

DOWN
1 Stock Cars have this type of engine
2 This is not allowed in NASCAR series racing
4 This type of cockpit allows the driver to get out quickly if there is a crash
5 Wheels that are outside the vehicle
8 NASCAR has than other series each year
10 Classic position for drivers in stock cars

Allows Bumping	Faster	Open Wheels
Bumping Allowed	Heavier	Oval Track
Cockpit Open	Less Races	Use Window
Driver Sits Center	Lighter	V Six Engine
Driver Sits Left	More Races	V Eight Engine
Enclosed Wheels	No Bumping	

Some Differences Between Race Cars Puzzle 7

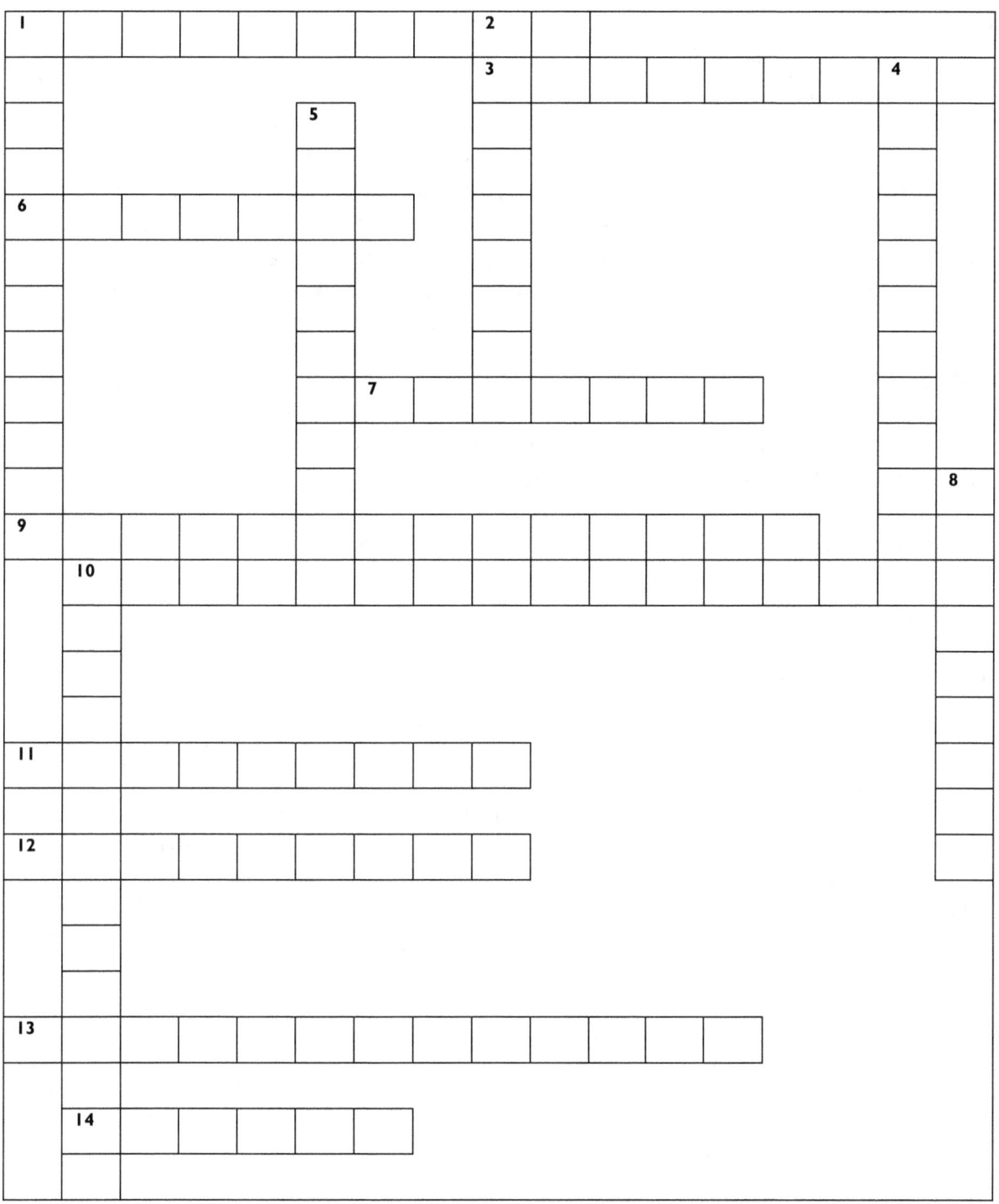

Notable Drivers Puzzle 8

ACROSS

1 Race car driver from Brazil, won the Indianapolis 500 four times
3 Won the first automobile race in America
4 Won two Daytona 500 races, he joined sports broadcasting as an analyst
8 Legendary driver inducted into the NASCAR Hall of Fame
9 Youngest woman to race in the Indianapolis 500
12 Had a reputation of an aggressive driver, nicknamed "The Intimidator"
14 Raced for 29 years and won the Indianapolis 500 four times
16 Raced for 16 straight seasons, many wins, one of NASCAR's finest drivers
17 Set a record by winning three consecutive Cup Series championships

DOWN

2 Third Generation driver won 1998 and 1999 NASCAR Xfinity Series Titles
5 Won races in NASCAR, IndyCar, World Sportscar Championship, Formula One
6 First woman to race in Indianapolis 500
7 Won seven NASCAR Championships
9 Held most wins in 1 season. Won most IndyCar Championships
10 Won Indianapolis 500 four times
11 Known for his success and wins in many types of racing
13 The King, most NASCAR wins
15 First woman to win an IndyCar Series Race

Andretti	Gordon	Unser
Castroneves	Guthrie	Wallace
Duryea	Johnson	Waltrip
Earnhardt	Mears	Yarborough
Earnhardt	Patrick	
Fisher	Pearson	
Foyt	Petty	

Notable Drivers Puzzle 8

Race Car Movies Puzzle 9

ACROSS

3 A movie about a Volkswagen Beetle Car
5 A movie about stock car racing starring Tom Cruise
8 A movie about a NASCAR driver starring Burt Reynolds
9 Movie about the 24 hour Le Mans race staring Steve McQueen
11 A kid's movie about Lightning McQueen

DOWN

1 The sheriff, bootleggers, and a fast driver give chase starring Burt Reynolds
2 A get away driver tries to quit when he meets a girl
4 The battle for the win between Ford and Ferrari at the Le Mans Race
6 A battle between NASCAR driver and Formula One driver
7 A series about racing staring Vin Diesel
10 A movie about four Grand Prix drivers

Baby Driver	Le Mans
Cars	Smokey And The Bandit
Days Of Thunder	Stroker Ace
Fast And Furious	Talladega Nights
Ford V Ferrari	The Love Bug
Grand Prix	

Race Car Movies Puzzle 9

ANSWERS

Types of Racing Puzzle 1

ACROSS
2   Stock Car Racing
5   Kart Racing
7   NASCAR
10 Indy
11 Drag Racing
13 Formula One
16 Hill Climb Racing

DOWN
1   Dirt Track Racing
3   Touring Car Racing
4   Grand Prix
6   Rally Cross
8   Sports Car Racing
9   Indy Car Series
12 Off Road Racing
14 Rallying
15 Formula E

											D			
	S	T	O	C	K	C	A	R	R	A	C	I	N	G
		O							R		R			
		U							T		A			
K	A	R	T	R	A	C	I	N	G		T		N	R
		I							R		D		A	
		N	A	S	C	A	R		S	A	P		L	
I		G						P	A	R		L		
N		C						O	K	I	N	D	Y	
D	R	A	G	R	A	C	I	N	G	R	X		C	
Y		R						T	R	A			R	
C		R						S	A	C			O	
A		A	O					C	I			S		
R		C	F	O	R	M	U	L	A	O	N	E		
S		I	F	A				R	G					
E		N	R	L				R						
R		G	O	L	F			A						
I			A	Y	O			C						
E			D	I	R			I						
S			R	N	M			N						
			A	G	U			G						
			C		L									
	H	I	L	L	C	L	I	M	B	R	A	C	I	N G
	N				A									
	G				E									

Races Puzzle 2

ACROSS
1 Cup Series
6 USAC
8 Xfinity Series
12 Indianapolis
13 NHRA

DOWN
2 Pikes Peak International
3 NASCAR Truck Series
4 Grand Prix
5 Le Mans Series
7 Indy Car
9 Formula One
10 Rally Cross
11 Daytona

Racing Facts Puzzle 3

ACROSS
3 Brickyard
4 Europe
6 Lollipop Man
7 Race On The Sand
8 A Pole Sitter
10 Bricks

13 Grand Prize
14 Nineteen Eleven

DOWN
1 A Bottle Of Milk
2 Qualifying Rounds
4 End Of The Race

5 Open Wheeled Cars
9 Starting Grid
11 Two Hundred
12 Cockpit

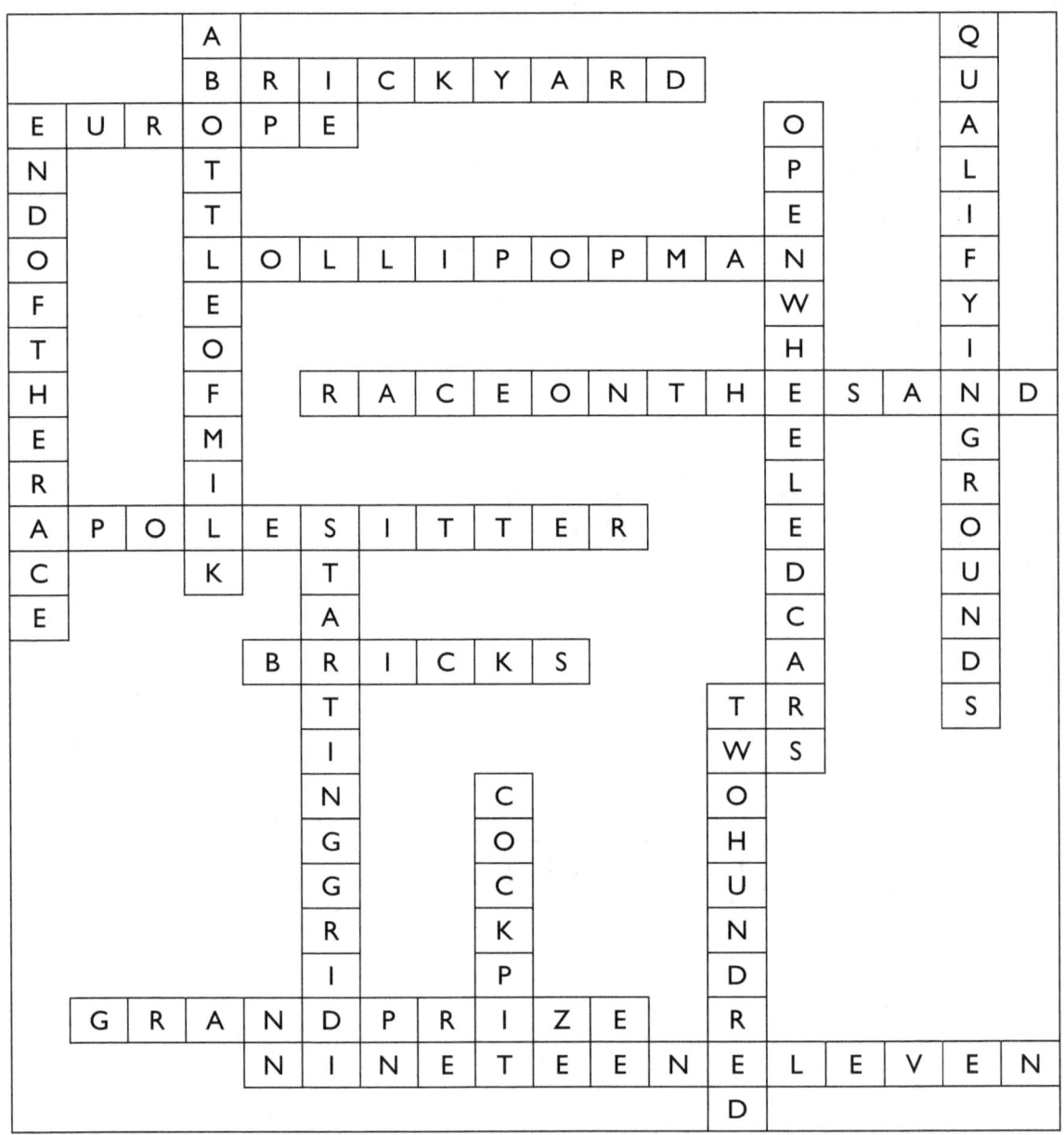

More Racing Facts Puzzle 4

ACROSS

1 Barriers
6 Yellow Flag
7 White Flag
8 Green Flag
9 Starting Grid
10 Victory Lane
11 Pit Stop
12 On The Pole
13 Black Flag
14 Wall Crash Prevent

DOWN

2 Red Flag
3 Eighteen Ninety Five
4 Checkered Flag
5 Blue Flag
7 William Randolph Hearst

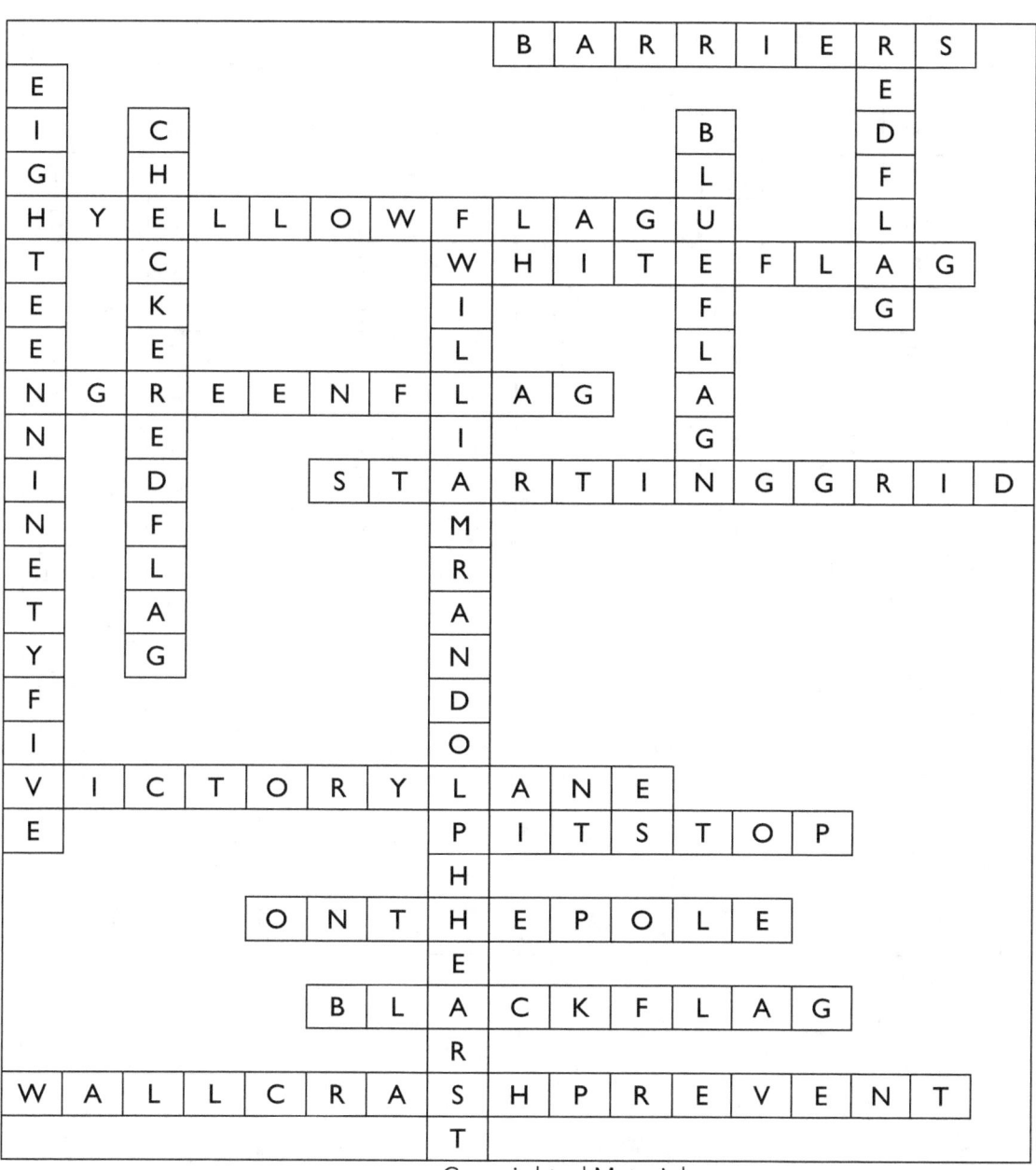

Parts of Race Cars Puzzle 5

ACROSS
2 Roll Cage
7 Wind Resistance
11 Airflow
14 Closed Wheel
16 Paddle Shift

DOWN
1 Fabricators
2 Restrictor Plates
3 Head Restraint
4 Closed Cockpit
5 Stock Car Window
6 Fuel Cell
7 Windshield

8 Thick Foam
9 Chassis
10 Frontal Protection
12 Open Cockpit
13 Inner Liner
15 Open Wheel

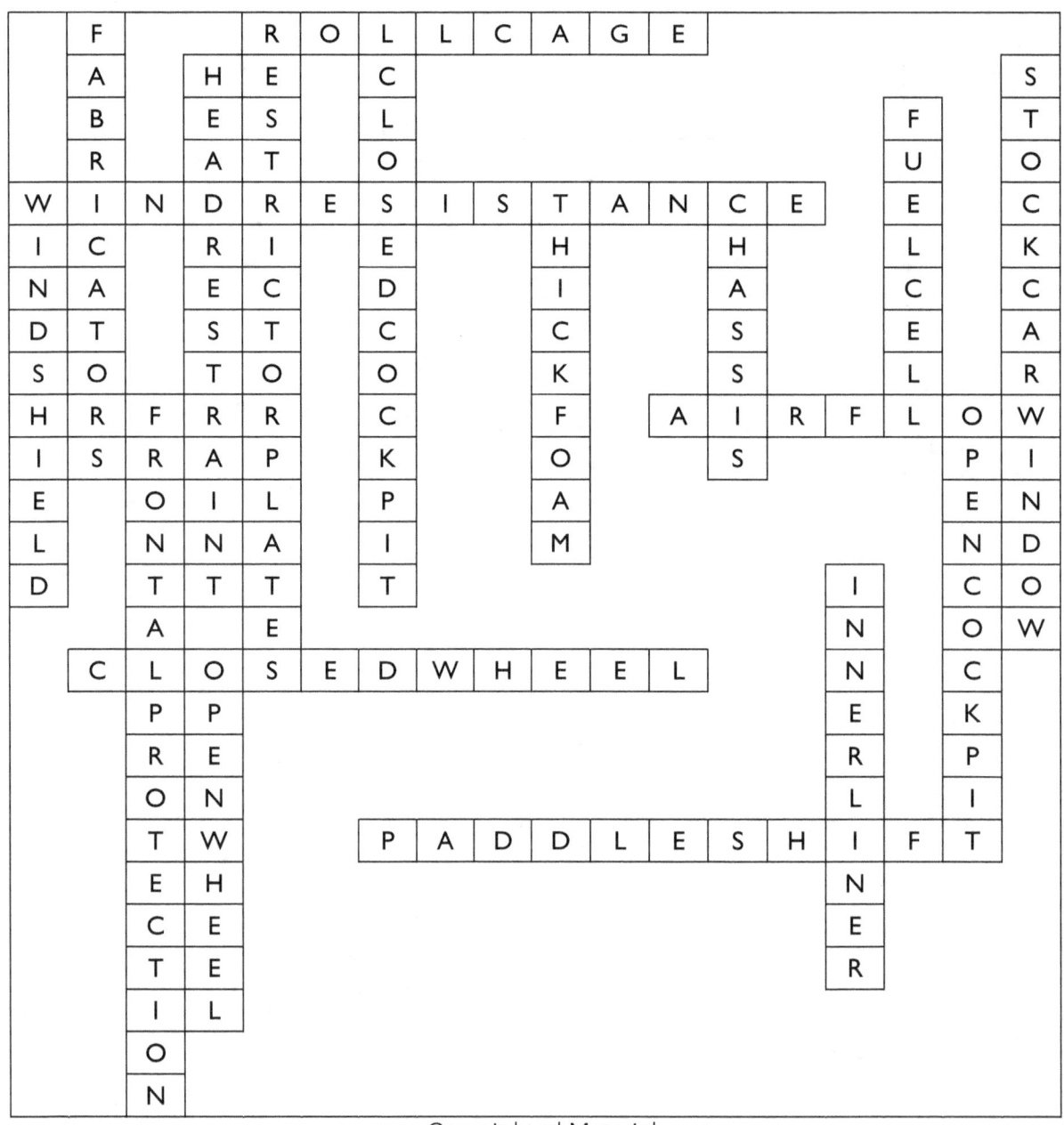

Pit Crew Puzzle 6

ACROSS
1 Speed Limit
4 Jackman
8 Crew Chief
10 Gym
11 Pit Stall
14 Engineers
15 Helmets
16 Tire Changers
17 Hand Speed
18 Hazards
19 Non Flamable

DOWN
1 Strategy
2 Catch Can
3 Hand Eye Coordination
5 Mechanics
6 Pit Box
7 Approved Safety Gear
9 Fills Fuel Tanks
10 Gloves
11 Pit Crew Training
12 Fire
13 Burned Hands

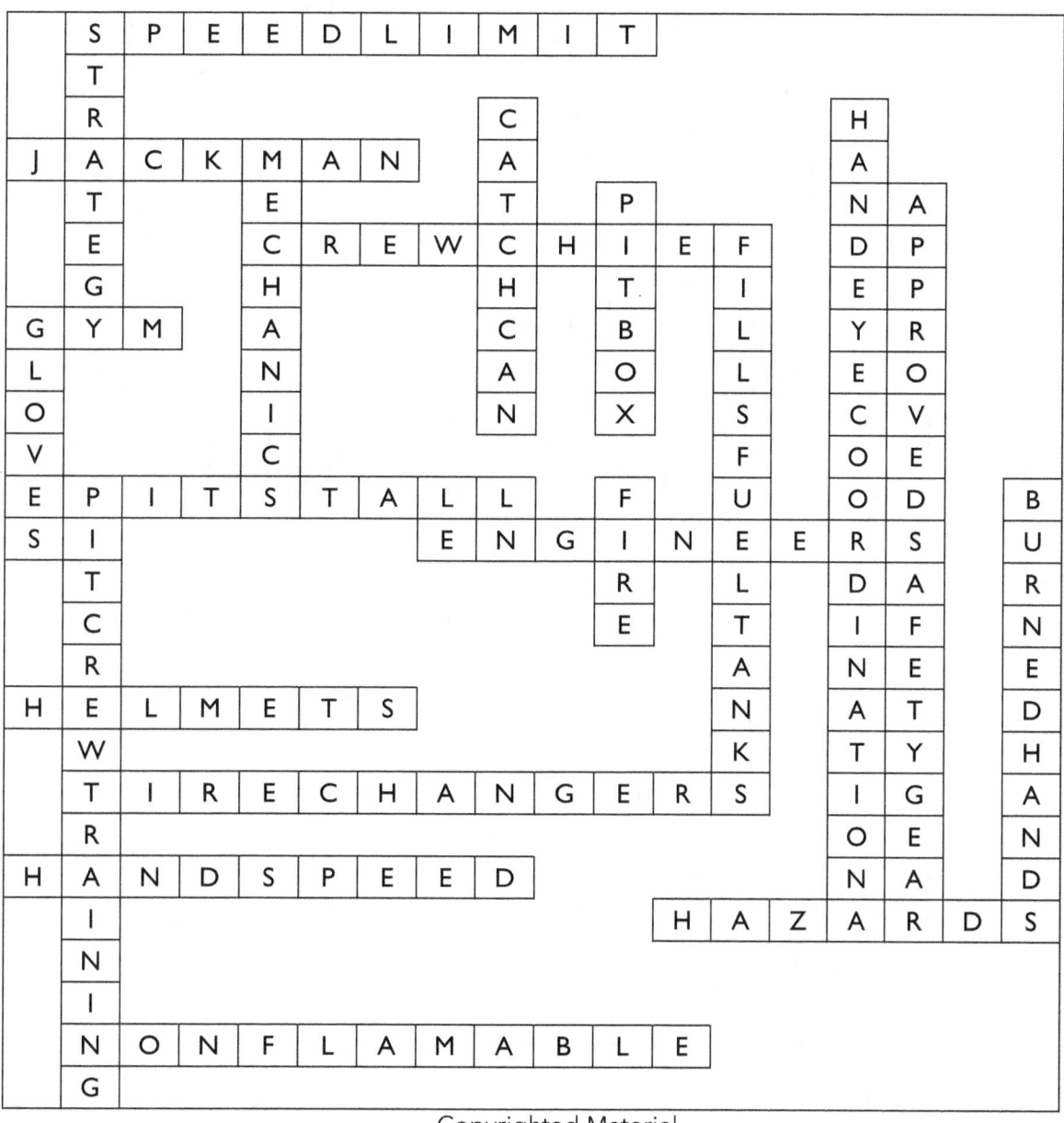

Some Differences Between Race Cars Puzzle 7

ACROSS
1 V Six Engine
3 Oval Track
6 Heavier
7 Lighter
9 Enclosed Wheels
10 Driver Sits Center

11 Less Races
12 Use Window
13 Allows Bumping
14 Faster

DOWN
1 V Eight Engine

2 No Bumping
4 Cockpit Open
5 Open Wheels
8 More Races
10 Driver Sits Left

V	S	I	X	E	N	G	I	N	E						
E								O	V	A	L	T	R	A	C	K
I				O				B						O	
G				P				U						C	
H	E	A	V	I	E	R		M						K	
T				N				P						P	
E				W				I						I	
N				H				N						T	
G				E	L	I	G	H	T	E	R			O	
I				E										P	
N				L										E	M
E	N	C	L	O	S	E	D	W	H	E	E	L	S	N	O
	D	R	I	V	E	R	S	I	T	S	C	E	N	T	E	R
	R														E
	I														R
	V														A
L	E	S	S	R	A	C	E	S							C
	R														E
U	S	E	W	I	N	D	O	W							S
	I													
	T													
	S													
A	L	L	O	W	S	B	U	M	P	I	N	G		
	E													
	F	A	S	T	E	R								
	T													

Notable Drivers Puzzle 8

ACROSS
1 Castroneves
3 Duryea
4 Waltrip
8 Gordon
9 Fisher
12 Earnhardt

14 Unser
16 Wallace
17 Yarborough

DOWN
2 Earnhardt
5 Andretti

6 Gutherie
7 Johnson
9 Foyt
10 Mears
11 Pearson
13 Petty
15 Patrick

Race Car Movies Puzzle 9

ACROSS

3 The Love Bug
5 Days Of Thunder
8 Stroker Ace
9 Le Mans
11 Cars

DOWN

1 Smokey And The Bandit
2 Baby Driver
4 Ford V Ferrari
6 Talladega Nights
7 Fast And Furious
10 Grand Prix

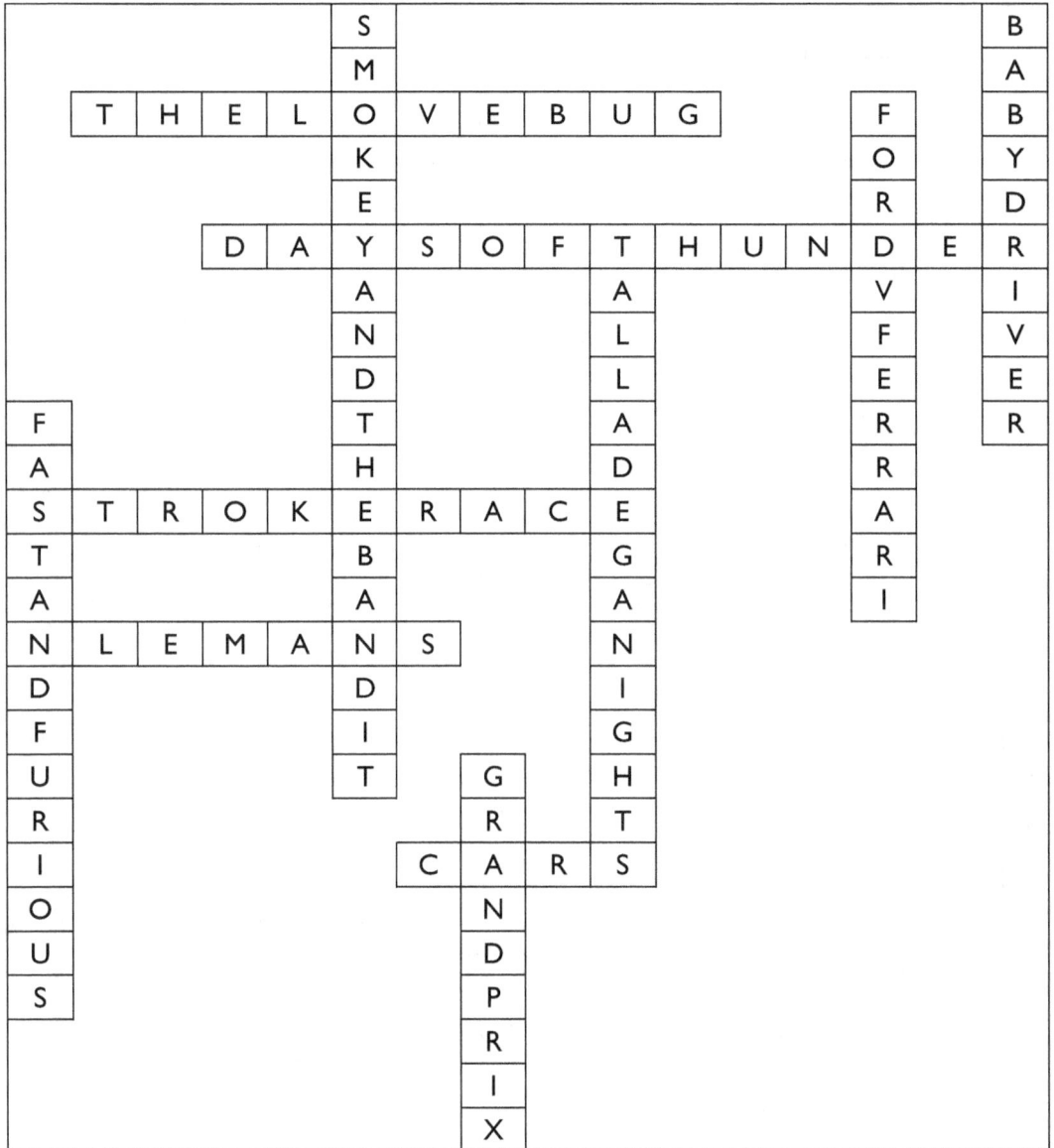

Sources

https://www.nascar.com   NASCAR History | Official Site Of NASCAR

https://www.racing-reference.info

https://motorsporttickets.com/blog/f1-pit-crew-what-each-member-does-during-a-formula-1-pit-stop/

https://www.chiff.com/articles/nascar-pit-crew.htm

https://www.Formula1.com

https://www.indycar.com

https://www.nascar.com

https://www.nascarkids.com

https://americanhistory.si.edu/america-on-the-move/essays/american-racing

https://www.motor.com/magazine-summary/racing-motor-early-years-march-2003/

https://www.rookieroad.com/nascar/what-is-nascar-stock-car/

https://KeepitGnarly.com/nascar-vs-indycar-what-is-the-difference/

https://bleacherreport.com/articles/1422027-the -25-greatest-american-drivers-of-all-time

https://www.merriam-webster.com/

Google searches

Wikipedia

Thank you for purchasing this Crossword Puzzle Book!!